Head Start to
A-Level Biology

A-Level Biology is a **big step up** from GCSE... no doubt about that.
But don't worry — this CGP book has been lovingly made to help you
hit the ground running at the start of your A-Level (or AS-Level) course.

It recaps everything you'll need to remember from GCSE, and introduces
some of the crucial concepts you'll meet at A-Level. For every topic, there are
crystal-clear study notes and plenty of **practice questions** to test your skills.

What CGP is all about

Our sole aim here at CGP is to produce the highest quality books
— carefully written, immaculately presented and dangerously
close to being funny.

Then we work our socks off to get them out to you
— at the cheapest possible prices.

Contents

Section 5 — Disease and Immunity

Section 6 — The Circulatory System

Section 7 — Variation, Evolution and Classification

Section 8 — Plants

Section 9 — Investigating and Interpreting

Published by CGP

Contributors:
Paddy Gannon
Barbara Green

Editors:
Christopher Lindle
Claire Plowman

ISBN: 978 1 78294 279 5

With thanks to Hayley Thompson for the proofreading.
With thanks to Laura Jakubowski for the copyright research.

Clipart from Corel®
Printed by Elanders Ltd, Newcastle upon Tyne.

Based on the classic CGP style created by Richard Parsons.

Proteins

Proteins are Made of Amino Acids

Proteins are composed of long chains of **amino acids**. There are **twenty different** amino acids used in proteins. They all contain carbon, hydrogen, oxygen and nitrogen, and some contain sulfur. All have the **same structure** as the one in the diagram but **R** can be one of twenty different chemical groups.

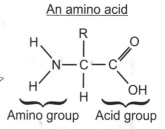

An amino acid

Amino group Acid group

Proteins are Held Together by Peptide Bonds

1) The chains of amino acids are attached to each other by **strong peptide bonds**.

2) The amino acids can be arranged in any sequence and proteins can be up to **several hundred** amino acids long.

3) The number of different proteins that are possible is almost unimaginable. Consider that there are several thousand ways of arranging a chain of just three amino acids, with each combination forming a different protein. Add one more amino acid to the chain and the number of possibilities leaps into the hundreds of thousands.

4) It's the **order** of the amino acids in a protein that determines its **structure** and it's the structure of a protein that determines **how it works**.

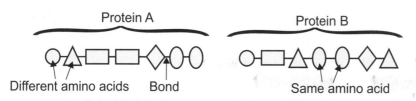

Protein A

Different amino acids Bond

Protein B

Same amino acid

(N.B. Each different shape represents a different amino acid.)

Each Protein has its Own Special Shape

1) The order in which the amino acids are arranged in a protein chain is called the **primary structure**.

2) Some chains **coil up** or **fold** into pleats that are held together by weak forces of chemical attraction called **hydrogen bonds**. The coils and pleats are the **secondary structure** of a protein.

3) Some proteins (especially enzymes) have a **tertiary structure**. The coiled chain of amino acids is folded into a **ball** that's held together by a mixture of weak chemical bonds (e.g. hydrogen bonds) and stronger bonds (e.g. disulfide bonds).

4) If the protein has a roughly spherical shape it's called a **globular protein** (e.g. enzymes are classed as globular proteins).

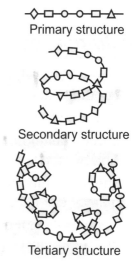

Primary structure

Secondary structure

Tertiary structure

The name's Bond. Peptide Bond...

1) What is the primary structure of a protein?

2) What type of bond holds together the secondary structure of a protein?

Carbohydrates

Carbohydrates Contain Three Elements

Carbohydrates contain **carbon**, **hydrogen** and **oxygen**. There are several types of carbohydrate, e.g. sugars, starch and cellulose.

1) Sugars are **small**, **water-soluble** molecules that taste sweet.

2) They're divided into two groups: **monosaccharides** (pronounced: mono-sack-a-rides) and **disaccharides** (die-sack-a-rides).

3) Monosaccharides are the **single units** from which all the other carbohydrates are built. **Glucose** and **fructose** are both monosaccharides. Glucose has two forms — **alpha** (α) and **beta** (β).

4) Disaccharides are formed when **two monosaccharides** are joined together by a chemical reaction. A molecule of **water** is also formed (so it's called a **condensation reaction**).

α-glucose molecule

ß-glucose molecule

The two forms of glucose have these groups swapped around.

GLUCOSE + GLUCOSE → MALTOSE (a disaccharide) + WATER

GLUCOSE + FRUCTOSE → SUCROSE (a disaccharide) + WATER

Starch is a Polysaccharide

Polysaccharides are **polymers** — large molecules made up of **monomers** (smaller units). The monomers of polysaccharides are **monosaccharides**. **Starch** molecules are made up of two different polysaccharides — **amylose** and **amylopectin**, which are polymers of glucose. The insoluble, compact starch molecules are an ideal way of **storing glucose**. Starch is **only** found in plant cells.

Amylose

Amylopectin

Cellulose is Also a Polysaccharide

1) Like starch, cellulose is a polymer of glucose, but the **bonding** between the glucose units is different.

2) As a result, the cellulose molecules are **long** and **straight**.

3) Several cellulose molecules can lie side by side to form **microfibrils**.

4) The molecules are held together by many weak **hydrogen bonds**.

5) Cellulose is only found in plant cells.

6) The microfibrils **strengthen** the plant cell wall.

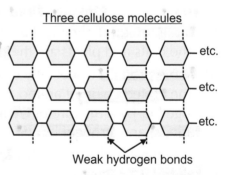

Three cellulose molecules

etc.

etc.

etc.

Weak hydrogen bonds

A poly-sack-a-ride — a bunch of kids on a helter skelter...

1) Name two monosaccharides.

2) Which disaccharide is composed of two molecules of glucose?

3) Name two polysaccharides.

Lipids

Lipids Contain *Carbon, Hydrogen* and *Oxygen*

Lipids are **oils** and **fats**. Plant oils and animal fats are mostly made up of a group of lipids called **triglycerides**. A triglyceride consists of a molecule of **glycerol** with **three fatty acids** attached to it.

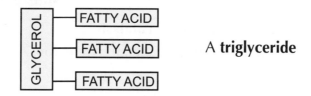

A **triglyceride**

A fatty acid molecule is a long chain of **carbon atoms** with an **acid group** (-COOH) at one end. **Hydrogen atoms** are attached to the carbon atoms. If every carbon atom in the chain is joined by a **single bond**, we say that the fatty acid is **saturated**. If one or more of the bonds is a **double bond**, it's said to be **unsaturated**. A fatty acid with many double bonds is **polyunsaturated**.

Saturated fatty acid

Unsaturated fatty acid

Phospholipids are a *Special* Type of *Lipid*

Phospholipids (pronounced: foss-foe-lip-id) are like triglycerides, but instead of having three fatty acid chains, they have **two** fatty acid chains and a **phosphate** group.
Cell membranes are made from a **double layer** of phospholipids.

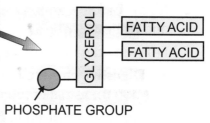

PHOSPHATE GROUP

Acid chain and the phospholipids — sounds like a punk band...

1) Which elements are fatty acids composed of?
2) What's the difference between saturated fatty acids and unsaturated fatty acids?
3) What's the difference between triglycerides and phospholipids?

Enzymes

Enzymes Help to *Speed up Biochemical Reactions*

1) In a living cell, thousands of **biochemical reactions** take place every second. The sum of these reactions is called **metabolism**. A single chain of these reactions is called a **metabolic pathway**.

2) Without enzymes, these reactions would take place very **slowly** at normal body temperature.

> 1) Enzymes are **biological catalysts**.
> 2) They **increase** the **rate** (speed) of reactions.

How do Enzymes *Act* as *Catalysts*?

1) Even reactions that release energy require an **input of energy** to get them going, e.g. the gas from a Bunsen burner doesn't burn until you provide heat energy from a match.

2) This input energy is called the **activation energy**. A reaction that needs a high activation energy can't start at a low temperature of 37 °C (i.e. body temperature).

3) Enzymes **reduce** the activation energy.

This graph shows the activation energies of a reaction **with** and **without** an enzyme:

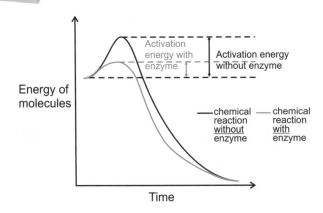

Enzymes are *Proteins*

1) All enzymes are **globular proteins** (because they're roughly spherical).

2) It's the order of amino acids in an enzyme that determines its **structure**, and so how it works.

3) Enzymes can be involved in **breaking down** molecules or **building** molecules. For example:

- **Digestive enzymes** are important in the digestive system, where they help to break down food into smaller molecules, e.g. carbohydrases break down carbohydrates.

- Enzymes involved in **DNA replication** help to build molecules, e.g. DNA polymerase.

I could really use a catalyst to help me write this gag...

1) What is the function of enzymes?

2) What is activation energy?

3) What do digestive enzymes do?

Enzymes

Enzymes have an **Active Site**

1) A substance that's acted upon by an enzyme is called its **substrate**.
2) The **active site** is a region on the surface of the enzyme molecule where a substrate molecule can attach itself. It's where the catalysed reaction takes place.
3) The shape of the substrate molecule and the shape of the active site are **complementary**, i.e. they fit each other.
4) Almost as soon as the **enzyme-substrate complex** has formed, the products of the reaction are released and the enzyme is ready to accept another substrate molecule.

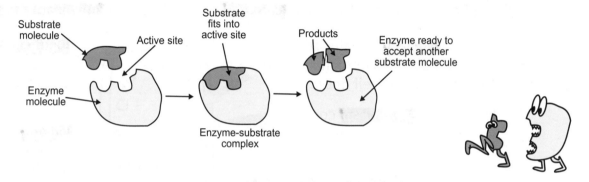

Enzymes are **Specific**

1) An enzyme usually catalyses one **specific** chemical reaction.
2) The substrate molecule must be the **correct shape** to fit into the active site.
3) **Only one substrate** will be the correct shape to fit, so each enzyme only catalyses one specific reaction.
4) Anything that **changes** the shape of the active site will **affect** how well the enzyme works.

The **Effect** of Temperature on **Enzyme Activity**

As temperature **increases**, enzyme reactions become **faster**, because the molecules have more **energy**. However, at high temperatures the atoms of the enzyme molecule vibrate more rapidly and **break** the weak bonds that hold the **tertiary structure** together. The **shape** of the active site **changes** and the substrate can no longer fit in. The enzyme is said to be **denatured**.

The **Effect** of pH on **Enzyme Activity**

Acids and **alkalis** can denature enzymes. Hydrogen ions (H^+) in acids and hydroxyl ions (OH^-) in alkalis disrupt the **weak bonds** and change the shape of the active site.

Lonely enzyme seeking complementary substrate...

1) Why are enzymes described as 'specific'?
2) Explain why a denatured enzyme will not function.
3) Describe the effect of pH on enzyme activity.

Eukaryotic and Prokaryotic Cells

Organisms can be Prokaryotes or Eukaryotes

1) **Prokaryotic** (pronounced like this: pro-carry-ot-ick) organisms are prokaryotic cells (i.e. they're **single-celled** organisms) and **eukaryotic** (you-carry-ot-ick) organisms are made up of eukaryotic cells.

2) Both types of cells contain **organelles**. Organelles are parts of cells — each one has a **specific function**.

Eukaryotic cells are **complex** and include all **animal** and **plant** cells. **Prokaryotic** cells are **smaller** and **simpler**, e.g. **bacteria**.

4 organelles **animal** and **plant** cells have in **common**:

Nucleus contains genetic material (DNA) that controls what the cell **does**.

Cytoplasm contains enzymes that speed up biochemical reactions.

Cell-surface membrane holds the cell together and controls what goes **in** and **out**.

Mitochondria are where glucose and oxygen are used in **respiration** to provide a source of **energy** for the cell.

3 extras that **only plant cells** have:

Rigid cell wall made of **cellulose**, gives the cell support.

Vacuole contains **cell sap**, a weak solution of sugar and salts.

Chloroplasts contain **chlorophyll** for **photosynthesis**. They're found in the **green** parts of plants, e.g. leaves and stem.

Bacterial Cells are Prokaryotic

1) Prokaryotes like bacteria are roughly a **tenth the size** of eukaryotic cells.

2) Prokaryotic cells **don't contain** a nucleus, mitochondria or chloroplasts.

3) As they **don't** have a nucleus, their **DNA floats freely** in the **cytoplasm**. Some prokaryotes also have **rings of DNA** called **plasmids**.

4) Some prokaryotes have a **flagellum** which **rotates** and allows the cell to **move**.

5) The diagram shows a bacterial cell as seen under an **electron microscope** (see next page).

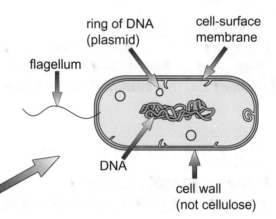

ring of DNA (plasmid)

cell-surface membrane

flagellum

DNA

cell wall (not cellulose)

Bacterial cheerleaders — they never stop swirling their flagella...

1) Give an example of a prokaryotic cell.

2) Name four organelles that plant and animals cells both have.

3) What is the function of mitochondria?

Microscopes

You Can **See Cell Structure** with a **Light Microscope**

A **light microscope** can magnify up to 1500 times and allows you to see individual animal and plant cells along with the organelles inside them.

1) If the cells have been **stained** you can see the dark-coloured **nucleus** surrounded by lighter-coloured **cytoplasm**.

2) Tiny **mitochondria** and the black line of the **cell membrane** are also visible.

3) In plant cells, the **cell wall**, **chloroplasts** and the **vacuole** can be seen.

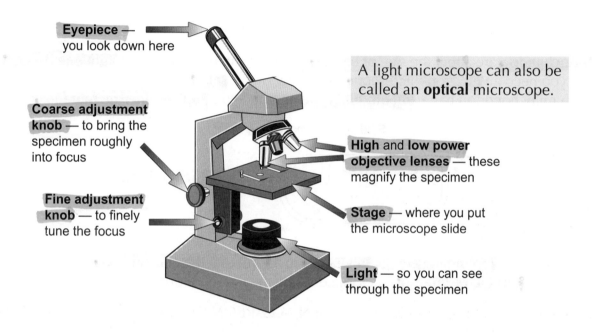

Eyepiece — you look down here

A light microscope can also be called an **optical** microscope.

Coarse adjustment knob — to bring the specimen roughly into focus

High and **low power objective lenses** — these magnify the specimen

Fine adjustment knob — to finely tune the focus

Stage — where you put the microscope slide

Light — so you can see through the specimen

Electron Microscopes have a **Greater Magnification**

1) The detailed **ultrastructure** of cells was revealed in the 1950s when the **electron microscope** was invented.

2) An electron microscope can **magnify** objects more than 500 000 times and, more importantly, it allows **greater detail** to be seen than a light microscope. For example, it allows you to see the detailed **structures inside organelles** such as mitochondria and chloroplasts.

3) The image that's recorded is called an **electron micrograph**.

I put a slide on the stage and then slid straight off the edge...

1) Name three things visible with a light microscope in both animal and plant cells.

2) Which type of microscope must be used to show the detailed ultrastructure of a cell?

3) What is the image recorded by an electron microscope called?

Functions of the Nucleus, Mitochondria and Cell Wall

Nucleus

1) The **nucleus** is the control centre of the cell.
2) It contains **DNA** (deoxyribonucleic acid): the coded information needed for **making proteins**.
3) During **cell division** the chromosomes carrying the long DNA molecules coil up, becoming shorter and thicker and visible with a light microscope.
4) Electron micrographs show that there's a **double membrane** around the nucleus.

Mitochondria

Mitochondria are about the size of bacteria, so they can be seen with a light microscope, but you need an electron microscope to see any of the detail.
Each mitochondrion has a **smooth outer membrane** and a **folded inner membrane**:

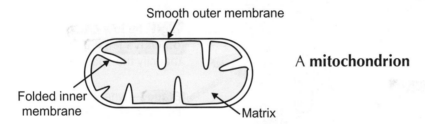

Smooth outer membrane

A **mitochondrion**

Folded inner membrane

Matrix

Their job is to capture the energy in glucose in a form that the cell can use.
To do this **aerobic respiration** takes place inside the mitochondria.

Word equation: GLUCOSE + OXYGEN → CARBON DIOXIDE + WATER + (ENERGY)

The energy released by respiration ends up in molecules of **ATP** (adenosine triphosphate).
ATP is used in the cell to provide the energy for **muscle contraction**, **active transport** (called active uptake in some text books) and **building large molecules** from small ones, as well as many other processes.

Cell Wall — Plants

1) The plant cell wall is relatively rigid and provides **support** for the cell.
2) It mainly consists of bundles of long, straight **cellulose molecules**.
3) The cellulose molecules lay side by side to form **microfibrils**.

Doctor, doctor my DNA is getting shorter and thicker...*

1) Which organelle acts as the control centre of the cell?
2) In which organelle does aerobic respiration occur?
3) Describe the membranes of a mitochondrion.
4) What is the word equation for aerobic respiration?
5) Name the molecule used to provide energy for processes in the cell.
6) Name the molecule that is found in bundles in plant cell walls.

Cell Membranes

Structure of the *Cell-Surface Membrane*

The **cell-surface membrane** is the very thin structure around an individual cell.

1) Electron micrographs show that the cell-surface membrane consists of a double layer of **phospholipid** molecules tightly packed together.

2) Bigger **protein molecules** are embedded in the phospholipid molecules.

3) Some proteins go **all the way through** the membrane and some only go **halfway**.

4) Membranes surrounding the **organelles** inside cells have the **same** structure.

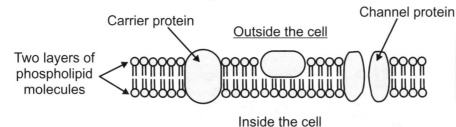

Carrier protein

Channel protein

Outside the cell

Two layers of
phospholipid
molecules

Inside the cell

Cell-surface membranes
can also be called
plasma membranes.

Do I **Really** *have to* **Know** *this Much* **Detail**?

1) The answer is "Yes". Once you're familiar with the molecular structure of the membrane you can explain how the membrane **controls** the passage of substances **in** and **out** of the cell.

2) Because the membrane only allows certain substances through it, it's described as being **partially permeable**.

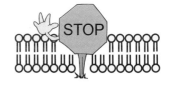

STOP

Substances **Pass Through** *Membranes by* **Four Methods**

1 **Diffusion**

1) The particles of liquids and gases are constantly **moving about**. This movement causes the particles to spread from an area of **higher** concentration to an area of **lower** concentration.

2) Particles will **diffuse** through the cell membrane as long as they are small enough to pass through the very small gaps **between** the phospholipid molecules. Water, oxygen and **carbon dioxide** molecules can do this.

3) The cell **doesn't** need to provide any energy for this process.

The difference in concentration is sometimes called a **concentration gradient**, e.g. a big difference in concentration is a big concentration gradient.

2 **Osmosis**

1) **Osmosis** is the diffusion of **water** molecules across a partially permeable membrane from a region of **higher concentration** of water molecules to a region of **lower concentration** of water molecules. The cell **doesn't** need to provide energy.

2) The concentration of water molecules is also referred to as the **water potential**. At AS and A-level, you tend to talk about water moving from a region of **higher water potential** to a region of **lower water potential**.

Cell Membranes

③ Facilitated Diffusion

1) Glucose and many other water soluble molecules are **too big** to diffuse across the membrane by themselves. They must be helped across by **carrier proteins**.

2) Each substance has its **own specific** carrier protein.

3) For example, a molecule of glucose fits onto the outside end of a **glucose carrier protein**.

4) This causes the protein to **change shape**, allowing the glucose molecule to diffuse through it into the cytoplasm of the cell. The cell **doesn't** need to provide any energy.

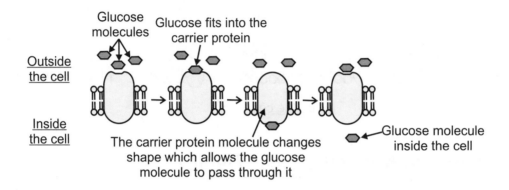

Glucose molecules

Glucose fits into the carrier protein

Outside the cell

Inside the cell

The carrier protein molecule changes shape which allows the glucose molecule to pass through it

Glucose molecule inside the cell

> **Mineral ions** like sodium (Na$^+$) and potassium (K$^+$) have electrical charges on them, so they also need help to cross the membrane. Specific **channel proteins** in the membrane allow them to diffuse through.

④ Active Transport (or Active Uptake)

1) When a cell needs to move substances across the membrane from a region of **low** concentration to a region of **higher** concentration, it must provide **energy**.

2) The substance fits into a **specific carrier protein**, then molecules of **ATP** (see page 8) provide the energy to change the shape of the protein.

3) As it changes shape the protein **actively transports** the substance across the membrane.

4) These special **carrier proteins** are sometimes called "**pumps**" because they're moving substances **against** a concentration gradient.

Active transport — isn't that just riding a bike?

1) Name the two types of molecule that make up the cell membrane.

2) Give four ways substances can cross cell membranes.

3) What do you call the diffusion of water molecules through the cell membrane?

4) Give another term for the concentration of water molecules.

5) Name the two types of protein involved in facilitated diffusion.

6) Why does active transport require ATP?

DNA and Protein Synthesis

DNA is Made Up of Nucleotides Containing Bases

1) DNA is a double helix (a double-stranded spiral). Each of the two DNA strands is made up of lots of small molecules called nucleotides.

2) Each nucleotide contains a part called a base. DNA has just four different bases.

3) These bases are: adenine (A), cytosine (C), guanine (G) and thymine (T).

4) Each base forms hydrogen bonds to a base on the other strand. This keeps the two DNA strands tightly wound together.

5) The bases always join up in the same way.

A DNA Double Helix

bases

base on one strand is joined to a base on the other by hydrogen bonds

strands

Adenine (A) always joins up with thymine (T), and cytosine (C) always joins up with guanine (G).

These pairs of bases are called complementary bases. They join together because they complement each other in shape — this is called complementary base pairing.

Proteins are Made by Reading the Code in DNA

1) DNA controls the production of proteins (protein synthesis) in a cell.

2) A section of DNA that codes for a particular protein is called a gene.

'Codes for' just means 'contains the instructions for'.

3) Proteins are made up of chains of amino acids. Each different protein has its own particular number and order of amino acids.

4) This gives each protein a different shape, which means each protein can have a different function.

5) It's the order of the bases in a gene that decides the order of amino acids in a protein.

6) Each gene contains a different sequence of bases — which is what allows it to code for a unique protein.

Pro-teen synthesis — supporting youth electronic music-making...

1) What is the name given to the double-stranded structure of DNA?

2) How many different bases are there in DNA?

3) Give the names of the bases in DNA.

4) How do the strands of DNA stay together?

5) What is complementary base pairing?

6) What is a gene?

7) What determines the order of amino acids in a protein?

RNA and Protein Synthesis

RNA is Needed to Make Proteins

1) DNA molecules (and so genes) are found in the **nucleus** of a cell, but they can't move out of the nucleus because they're very **large**.

2) Protein synthesis happens in the **cytoplasm** at organelles called **ribosomes**.

3) So when a cell **needs** a particular protein, a **copy** of the gene that codes for it is made in the nucleus. This copy is **smaller** than DNA so it can move in to the cytoplasm, where it can be used to make the protein.

4) The copy of the gene is made from a molecule called **messenger RNA** (mRNA).

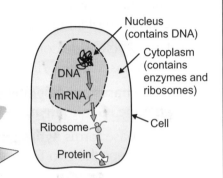

DNA is Used as a Template to Make an mRNA Molecule

1) The DNA in the gene acts as a **template**.

2) RNA, like DNA, is made up of **nucleotides**, which each have a **base**.

3) The bases on RNA nucleotides line up next to their **complementary** bases on the DNA template.

- In RNA, there's **no thymine (T)**, so the base **uracil (U)** binds to any **adenine (A)** in the DNA instead.

- Once the bases on the **RNA** nucleotides have **paired up** with the bases on the **DNA** strand, the RNA nucleotides join together to make an **mRNA molecule**.

4) Eventually, a **whole copy** of the gene is made and the **sequence** (order) of **bases** in the mRNA copy is complementary to the sequence of bases in the DNA template.

Complementary base pairs in DNA

A + T
C + G

Complementary base pairs in RNA

A + U
C + G

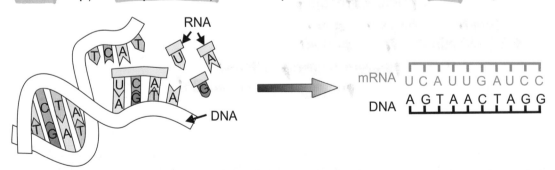

mRNA U C A U U G A U C C
DNA A G T A A C T A G G

Complimentary RNA — oh, you do look dashing Mr Ribo Some...

1) Why does a copy of a gene need to be made for protein synthesis?

2) What does the 'm' in mRNA stand for?

3) In RNA, which base is complementary to adenine?

4) Give the mRNA sequence that would be complementary to the DNA sequence: ATTGCGCA

Mutations

*The **Order** of **Bases** Determines the **Order** of **Amino Acids***

Three bases in a row (a **triplet**, e.g. GCT) codes for **one amino acid** — this is called the **genetic code**. **Different amino acids** are coded for by **different triplets**, e.g. TAT = tyrosine, AGT = serine. The **order of the bases** (and so triplets) in the DNA of a gene determines the order of bases in its mRNA copy, and that determines the **order of amino acids** in a protein:

***Mutations** Change the **Order of Bases** in DNA*

1) **Mutations** are changes to the **base** sequence (order) of DNA.

2) For example, one base can be **substituted** (swapped) for another one. This can cause the base triplet to **change**. E.g. if C is substituted for A, GCT becomes GAT.

3) So mutations can change the **amino acids** in the protein that the gene codes for.

4) A change in the amino acids can cause a **different protein** to be produced. Sometimes the different protein can be **harmful** (see below).

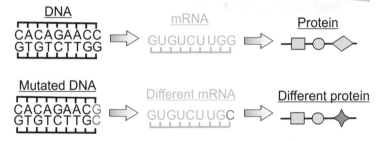

Mutations happen **spontaneously** (randomly), but how **frequently** they happen can be increased by **mutagenic agents** — factors that increase mutations, e.g. UV radiation in sunlight.

Mutations** can be **Harmful

1) Mutations can cause **cancer** because **cell division** is controlled by **proteins**. If mutations occur in the **genes** for these proteins, they can **alter** the proteins so they **no longer work**. This can lead to **uncontrolled cell division**, and the development of a **tumour** (cancer).

2) Mutations also cause **genetic disorders** — mutations that result in **altered** genes and proteins can be **inherited** (passed on from your parents), e.g. cystic fibrosis.

DoNAtello, LeAmino... it's the Teenage Mutant Protein Makers...

1) How many bases code for one amino acid?

2) What are mutations?

3) What do mutagenic agents do?

Chromosomes

DNA is Found on Chromosomes

DNA is found in the **nucleus** of **eukaryotic cells**. It has to be **wound up** into chromosomes to fit in. Each human chromosome contains between a couple of hundred and a few thousand genes.

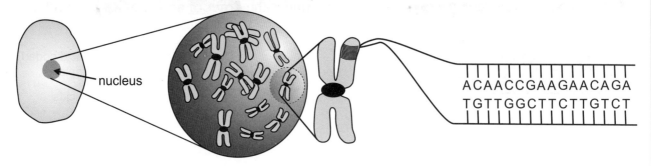

nucleus

ACAACCGAAGAACAGA
TGTTGGCTTCTTGTCT

Homologous Pairs

Humans have **23 homologous pairs** of chromosomes (46 in total), e.g. two number 1s, two number 2s, two number 3s, etc. One from each pair comes from your mother and one comes from your father. Both chromosomes in a pair are the **same size** and carry the **same genes** (which is why they're called **homologous pairs**). But they usually have **different alleles** (different versions of the genes).

Chromosomes are Often Shown as X-Shaped

In loads of books chromosomes are shown as **X-shaped**. An X-shaped chromosome is actually **one chromosome** attached to an **identical copy** of itself. Don't get it confused with a homologous pair of chromosomes. They're only X-shaped just after the DNA has been **replicated** (e.g. in cell division). Each side of the X is referred to as a **chromatid** and the bit in the middle where they're attached is called the **centromere**.

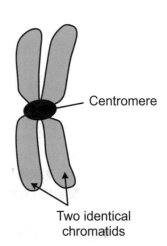

Centromere

Two identical chromatids

It's in his DNA, D, D, D, DNA...

1) Where is DNA found in a eukaryotic cell?
2) How many homologous pairs of chromosomes do human cells have?
3) Are homologous pairs of chromosomes identical? Explain your answer.
4) What is a chromatid?
5) What is the name of the region where two identical chromatids are joined?

Cell Division — Mitosis

Mitosis is Needed for Growth and Repair

1) If you have **damaged** tissue, the cells around the damaged area divide by **mitosis** to replace the damaged cells.
2) Cells **also** divide by mitosis to produce new tissue for **growth**.

Asexual Reproduction Involves Mitosis

1) In **asexual reproduction**, a single organism produces offspring by dividing into two organisms or by splitting off a piece of itself.
2) All the offspring are **genetically identical** to each other and to the parent.
3) The cells divide by **mitosis** (like most cells).

Bacteria and many plants reproduce asexually.

In Mitosis the DNA Copies Itself, Then the Cell Divides Once

Mitosis is split up into **four** stages: **prophase**, **metaphase**, **anaphase** and **telophase**. **Before** mitosis starts, there's a period called **interphase**.

1) **Interphase** — Before the cell starts to divide, every DNA molecule (each chromosome) must **replicate** so that each new cell has a full copy of DNA. The new molecule remains attached to the original one at the **centromere**.

2) **Prophase** — Mitosis can now begin. Each DNA molecule becomes **supercoiled** and **compact**. Each chromosome can now be seen with a light microscope and appears as **two chromatids** lying side by side, joined by the centromere (i.e. X-shaped).

3) **Metaphase** — The **nuclear membrane** breaks down and the chromosomes **line up** along the **equator** (middle) of the cell.

4) **Anaphase** — The centromeres split and the **chromatids separate** and are dragged to **opposite** ends of the cell.

5) **Telophase** — A **nuclear membrane** forms around each set of chromatids (exact copies of the original chromosomes) and the **cytoplasm divides**.

MITOSIS

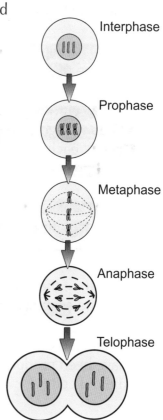

Interphase

Prophase

Metaphase

Anaphase

Telophase

Ouch, you stepped on my toe, sis... ba dum tsh

1) Give three uses of mitosis.
2) Why is DNA replicated before cell division can occur?
3) Do the homologous pairs separate in mitosis?
4) How many cells are produced when a cell divides by mitosis?

Cell Division — Meiosis

Sexual Reproduction Involves Meiosis

1) In **sexual reproduction**, the offspring are genetically different from their parents and from each other. This produces variation in a population.

2) Each parent produces sex cells (**gametes**) containing just **one set** of genetic material. This involves a special kind of cell division, called **meiosis**, and the gametes are described as being **haploid**.

3) During **fertilisation** the nuclei of the gametes join together to form a **zygote**. The zygote has **two complete sets** of genetic material, and is said to be **diploid**.

4) The zygote grows by simple cell division (**mitosis**) to form the **embryo**.

In Meiosis, DNA Copies Itself Then the Cell Divides Twice

1) The **only cells** in the human body that divide by meiosis are special cells in the **testes** and **ovaries**.

2) These cells divide to produce **gametes** (sperm and eggs).

3) The DNA **replicates**, so each of the 46 chromosomes become two chromatids joined by a centromere.

4) The 46 chromosomes sort themselves into the **23 homologous pairs**, then the **pairs separate**. One of each pair goes to one side of the cell and one goes to the other.

5) The cytoplasm now divides. Each of the new cells **contains 23 chromosomes** (consisting of two chromatids joined by a centromere).

6) In both of these new cells the **chromatids separate** and the cytoplasm divides to form two cells.

7) At the end of meiosis, **four haploid cells** have been produced from every original diploid cell.

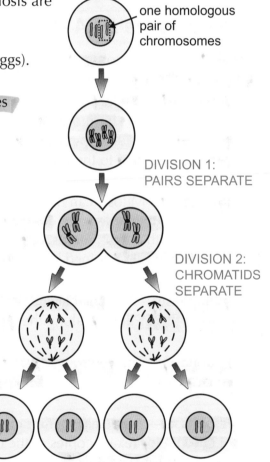

one homologous pair of chromosomes

DIVISION 1: PAIRS SEPARATE

DIVISION 2: CHROMATIDS SEPARATE

Worry not, before you know it your A-level testes will soon be ovaries...

1) Are gametes haploid or diploid?
2) Where in the human body does meiosis occur?
3) How many cell divisions are there in meiosis?
4) How many cells are produced when a cell divides by meiosis?

Size and Surface Area to Volume Ratio

Small Objects have Relatively *Large Surface Areas*

1) Have you ever wondered **why** there are no large single-celled organisms or why big animals are made up of **millions** of tiny cells instead of a few large ones?

2) The main reason relates to the changes in the **surface area to volume ratio** of an object as it increases in size.

3) Look at the three cubes in the diagram below. The **smallest cube** has the **biggest** surface area to volume ratio and the **biggest cube** has the **smallest** surface area to volume ratio.

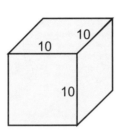
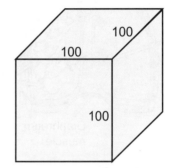

Surface area = 6 cm²
Volume = 1 cm³
Surface area : Volume
6 : 1

Surface area = 600 cm²
Volume = 1000 cm³
Surface area : Volume
0.6 : 1

Surface area = 60,000 cm²
Volume = 1,000,000 cm³
Surface area : Volume
0.06 : 1

Surface Area is Important for *Exchange*

1) Cells or organisms need to **exchange materials** and **heat** with their environment.

2) **More** chemical reactions happen every second in organisms with a **larger volume** than in ones with smaller volumes.

3) Therefore **more** oxygen, nutrients, waste products and heat need to be exchanged across the membrane of cells of larger organisms.

4) With increasing volume this becomes an **ever-increasing problem**.

My surface area just keeps growing... so does my volume (it's the pies)...

1) Which has the bigger surface area to volume ratio, a small organism or a large organism?

2) An animal has a surface area of 7.5 cm² and a volume of 1 cm³. What is its surface area to volume ratio?

3) Which animal has the greatest surface area to volume ratio — Animal A (9.8 : 1), Animal B (0.98 : 1)?

4) Give three materials that need to be exchanged across the membranes of organisms' cells.

Structure of the Thorax

Lungs have a Very Large Gas Exchange Surface

Large, active animals, like mammals, have evolved complex **blood systems** and **lungs** to provide a **large surface area** for the efficient diffusion of oxygen and carbon dioxide.

Gas exchange takes place in millions of tiny air sacs, called **alveoli**.

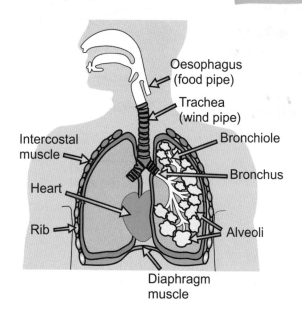

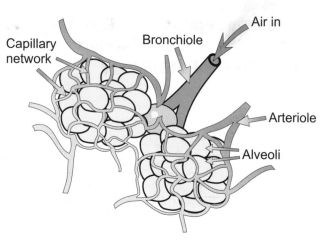

Alveoli have Adaptations that Increase the Diffusion Rate

1) The walls of the alveoli consist of a **single layer** of thin, flattened, epithelial cells. Diffusion happens **faster** when molecules only have to travel **short** distances.

2) Diffusion is faster when there's a **bigger difference** in concentrations between two regions. The blood flowing through the rich network of capillaries around the alveoli **carries away** the oxygen that has diffused through the alveolar walls. This ensures that there's always a **higher concentration of oxygen** inside the alveoli than in the blood. The reverse is true for **carbon dioxide**.

3) The alveolar walls are **fully permeable** to dissolved gases. Oxygen and carbon dioxide can pass easily through the cell membranes of the epithelial cells.

I like my ralveoli filled with spinach and ricotta...

1) Why have large mammals evolved complex blood systems and lungs?

2) In which part of the lungs does gas exchange take place?

3) Describe the shape of the cells that make up the walls of the alveoli and explain how their shape suits their function.

4) What type of cell are the alveoli walls made of?

5) a) Why does oxygen diffuse from inside the alveoli into the blood?
   b) Name another gas that can pass easily through the walls of the alveoli.

Breathing In and Breathing Out

Why do We Need to Breathe?

Ventilation (breathing) ensures that air with a **high concentration of oxygen** is taken into the lungs, and air with a **high concentration of carbon dioxide** is removed from the lungs. This maintains high **concentration gradients** between air (inside your alveoli) and blood, **increasing** the rate of diffusion of oxygen and carbon dioxide.

If Volume Increases, Air Pressure Decreases

If the **volume** of an enclosed space is **increased**, the **pressure** inside it will **decrease**.

1) The lungs are suspended in the **airtight thorax**.

2) Increasing the volume of the thorax decreases the air pressure in the lungs to below atmospheric pressure. Air flows **into** the lungs, inflating them until the pressure in the alveoli equals that of the atmosphere.

3) Decreasing the volume of the thorax increases the pressure in the lungs and air **flows out** until the pressure in the alveoli drops to atmospheric pressure.

Breathing In...

1) **Intercostal muscles** and **diaphragm** (a muscular sheet) **contract**.
2) Thorax volume **increases**.
3) This decreases the pressure, so air **flows in**.

...and Breathing Out

1) **Intercostal muscles** and **diaphragm relax**.
2) Thorax volume **decreases**.
3) This increases the pressure, so air flows **out**.

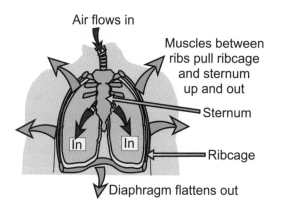

Air flows in

Muscles between ribs pull ribcage and sternum up and out

Sternum

In | In

Ribcage

Diaphragm flattens out

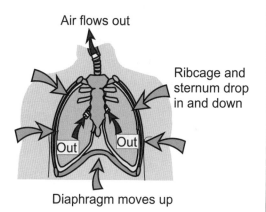

Air flows out

Ribcage and sternum drop in and down

Out | Out

Diaphragm moves up

We're heading inter-costal waters — look out for rocks at the sternum...

1) Describe the relationship between volume and pressure in an enclosed space.
2) Does the volume of the thorax increase or decrease when you breathe out?
3) Which two sets of muscles contract when we breathe in?

Disease

Disease can be Caused by Many Things

1) **Pathogens** — these are organisms that can cause disease, e.g. bacteria and viruses. **Infectious diseases** are caused by pathogens and can be passed from person to person, e.g. TB, malaria and HIV.

2) **Genetic defects** — some diseases are caused by **mutations** in a person's genes, e.g. cystic fibrosis is caused by a mutation in a gene for a protein.

3) **Lifestyle** — certain lifestyles **increase the risk** of getting some diseases, e.g. smokers are more likely to get lung cancer.

Risk Factors for Disease

1) A risk factor is something that **increases the chances** of something bad happening. For example, smoking is a risk factor for heart disease — if you smoke you're **more likely** to get heart disease.

2) Risk factors **don't always** lead to disease though. For example, using sunbeds is a risk factor for skin cancer — if you use sunbeds you increase your risk of skin cancer, but you won't necessarily get the disease.

3) Some risk factors are **unavoidable** because they're **inherited**, e.g. certain versions of genes increase your risk of getting breast cancer.

4) Some risk factors are **avoidable** because they're associated with your **lifestyle**. For example, a diet high in salt is a risk factor for high blood pressure — if you change your lifestyle to reduce your salt intake you reduce the risk.

Here's a table showing some common **lifestyle** risk factors and the diseases they're associated with:

Risk factor	Diseases
Smoking	Mouth, lung and throat cancer, emphysema and other lung diseases, cardiovascular disease
Drinking too much alcohol	Mouth, stomach, liver and breast cancer, possibly many other cancers, cardiovascular disease
High blood pressure	Cardiovascular disease, diabetes
Overweight/obese	Various cancers, cardiovascular disease, diabetes
Unbalanced diet	Various cancers, cardiovascular disease, diabetes
Using sun beds too much	Skin cancer

Taking your Nan's fashion advice — a risk factor for embarrassment...

1) What are pathogens?

2) Give an example of an infectious disease.

3) What is a risk factor?

4) List two diseases that smoking is a risk factor for.

Immunity

Phagocytes Engulf Pathogens

1) If a pathogen gets into the body it's detected by a type of white blood cell called a **phagocyte**.

2) It's actually the **molecules** on the **surface** of the pathogen that the phagocytes detect. These molecules are called **antigens**.

3) Human cells have antigens on their surface too, but phagocytes can tell the difference between '**self**' (your own) and '**foreign**' antigens.

There are lots of **different types** of white blood cells.

4) Phagocytes **engulf** pathogens that are carrying foreign antigens and destroy them.

White Blood Cells Produce Antibodies

1) Some white blood cells produce **antibodies** that **bind to** antigens.

2) The ones that produce antibodies are called **B-cells** (they're sometimes called B-lymphocytes — pronounced: lim-fo-sites).

3) When the antibody binds to the antigen it brings about the **death** of the pathogen carrying it.

Another Type of White Blood Cell is Involved

1) **T-cells** (or T-lymphocytes) are a type of white blood cell that are involved in **communication** between phagocytes and B-cells.

2) When a phagocyte has engulfed a pathogen it signals to the T-cell that it's found something. The T-cell then **activates** the B-cells to produce antibodies.

Vaccination Gives You Immunity

1) If you're **vaccinated** against a pathogen you can't get that disease (you're **immune**).

2) Vaccines **contain antigens** from a pathogen in a form that can't harm you, e.g. attached to dead bacteria.

3) Your body produces **antibodies** against the antigens so, if the same pathogen (carrying the same antigens) tries to invade again, the immune system can respond **really quickly** and you won't suffer from any **symptoms**.

4) Vaccines don't stop the pathogen getting **into** the body, they just **get rid of it** really quickly when it does.

I seem to be immune to learning all this Biology...

1) What do phagocytes detect?

2) What kind of white blood cells produce antibodies?

3) What is the role of T-cells?

4) What do vaccines contain?

The Circulatory System

Large Animals Need a Circulatory System

1) Diffusion is only efficient over **short distances**, so any animal bigger than a simple worm needs a system that will bring glucose and oxygen into close contact with individual cells.

2) In humans, the **heart** pumps blood around the body through **blood vessels**.

The heart has **four chambers** — two **atria** and two **ventricles**.

The **heart**

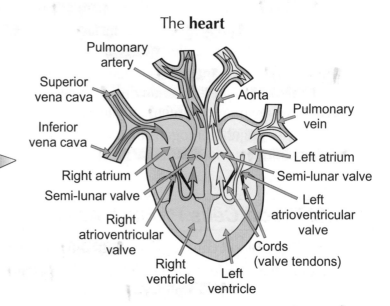

3) The blood vessels carry blood around the **entire body** and go to **every organ** before returning the blood to the heart.

4) There are **three** main types of blood vessel:
   - **Arteries** carry blood **away** from the heart.
   - **Veins** carry blood **to** the heart.
   - **Capillaries** are where the exchange between the blood and the cells takes place.

5) As the blood flows through the **tissues**, dissolved substances such as glucose, oxygen and carbon dioxide are **exchanged** between the blood and the cells.

The main artery in the human body is the **aorta**. It carries oxygenated blood **from the heart** to the rest of the body.

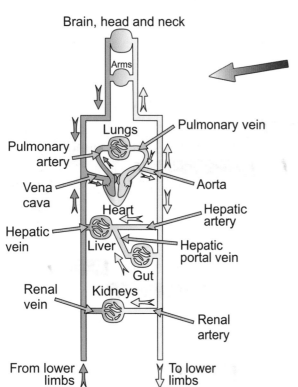

A circulatory system — going round and round the M25...

1) Name the organ that pumps blood around the body.
2) Name the four chambers of the heart.
3) Name the three main types of blood vessel.
4) In which type of blood vessel are substances exchanged between the blood and the cells?

The Heart

Important *Facts* to *Remember*

1) The heart acts like two separate **pumps**. The **right side** sends blood to the **lungs** and the **left side** pumps blood around the rest of the **body**.

2) Blood always flows from a region of **higher pressure** to a region of **lower pressure**.

3) **Valves** in the heart prevent the blood from flowing backwards.

4) **No energy** is required to make the valves work — it's the **blood pressing** on the valves that makes them **open and close**.

The *Cardiac Cycle*

The **cardiac cycle** is the sequence of events that occurs during **one heartbeat**.

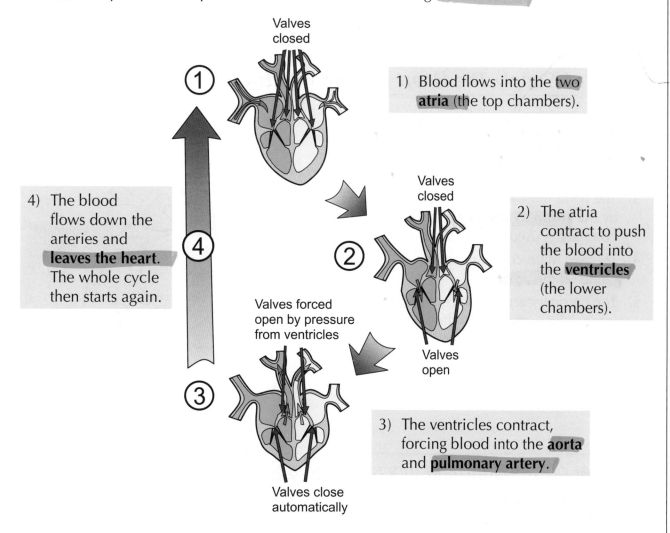

Valves closed

1) Blood flows into the two **atria** (the top chambers).

Valves closed

2) The atria contract to push the blood into the **ventricles** (the lower chambers).

Valves open

Valves forced open by pressure from ventricles

4) The blood flows down the arteries and **leaves the heart**. The whole cycle then starts again.

3) The ventricles contract, forcing blood into the **aorta** and **pulmonary artery**.

Valves close automatically

The **ventricles** are much more **powerful** than the atria and, when they contract, the **heart valves** pop shut automatically to prevent **backflow** into the atria. The ventricle walls are **thicker** because they need to push the blood further (e.g. the **left ventricle** has to push blood all the way round the body).

As soon as the ventricles relax, the valves at the top of the heart **pop shut** to prevent backflow of blood (back into the ventricles) as the blood in the arteries is now under **a fair bit of pressure**.

The Heart

The **Heart** has its Own **Pacemaker**

1) **Most muscles** require **nerve impulses** from the central nervous system to make them **contract**.

2) The heart **produces** its own **electrical impulses**.

3) A group of specialised cells called the **sino-atrial node**, in the wall of the right atrium, sends out **regular impulses**.

4) These spread across the atria, making them **contract**.

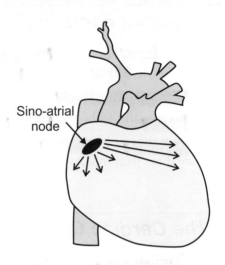

Sino-atrial node

Arteries Supply The **Heart Muscle** with **Blood**

1) Heart muscle, like all tissue, needs **oxygen** and **glucose** so it can respire and release the energy it needs to function.

2) It gets these things from the **blood**.

3) The heart muscle is supplied with blood by the **coronary arteries** (the word coronary is used to refer to the heart).

4) The two main coronary arteries come off the **aorta**.

5) The coronary arteries are quite **thin** (especially compared to the aorta).

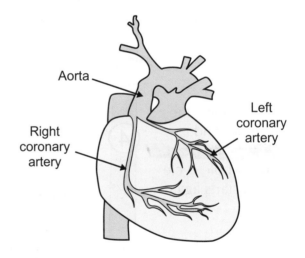

Aorta

Right coronary artery

Left coronary artery

Aorta get on with learning this stuff, I suppose...

1) Does the right hand side of the heart pump blood to the body or to the lungs?

2) What is the function of the heart valves?

3) Do heart valves require energy to open and close?

4) Where does the blood go after leaving the atria?

5) Why are the walls of the ventricles thicker than the walls of the atria?

6) The sino-atrial node is sometimes called the heart's natural pacemaker. What is its function?

7) Why does heart muscle require a blood supply?

8) Name the blood vessels that supply the heart muscle with blood.

Blood Vessels

Arteries, Arterioles, Capillaries and Veins

1) **Arteries** carry blood away from the heart.
2) They subdivide into smaller vessels called **arterioles**.
3) Arterioles subdivide into microscopic vessels called **capillaries**.
4) Capillaries join up to form **veins**.
5) Veins **return** blood to the **heart**.

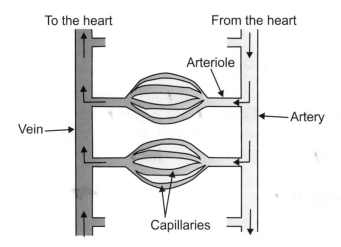

Arteries are Elastic

1) Arteries have a **thick wall** compared to the diameter of the lumen. There's an outer layer of **fibrous tissue**, then a thick layer of **elastic tissue** and **smooth muscle**, then a very thin inner layer of folded **endothelial tissue**.

2) When the ventricles contract, blood enters the arteries at **high pressure**. This **stretches** the folded endothelium and elastic walls. When the ventricles relax, it's the elastic recoil of the artery wall (when the wall shrinks back to its original size) that keeps the blood pressure up. Important organs, like the kidneys, wouldn't be able to function if the blood pressure dropped too far between heartbeats.

Elastic tissue in wall

Thick muscle layer

Lumen (hole in centre, through which blood flows)

Folded endothelium

Arterioles can Contract

1) Arterioles are **narrower** than arteries and they have a higher proportion of **smooth muscle fibres** and a lower proportion of **elastic tissue**.

Mainly circular muscle

2) When the circular muscle fibres of an arteriole contract, the diameter of the lumen is reduced, so **less blood flows** through that vessel. This means that arterioles can **control** the amount of blood flowing to a particular organ.

Blood Vessels

Capillaries can Only be Seen With a Microscope

Capillary walls consist of a single layer of **endothelial cells** (cells that line the blood vessels). Some capillaries have **tiny gaps** between the endothelial cells.

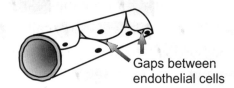

Gaps between endothelial cells

Capillaries are Well Suited to Their Job

1) The **very thin walls** and the **gaps** between the cells allow water and substances like glucose and oxygen to **diffuse quickly** from the blood into the cells. Waste products, such as carbon dioxide and urea, diffuse from the cells into the blood.

2) Organs contain thousands of capillaries, so altogether there's a **huge surface area** for the exchange of substances.

3) Blood flows quite **slowly** through capillaries. This allows **more time** for diffusion to occur.

Veins Have Valves

1) A vein has a **large lumen** and a relatively thin wall containing some elastic tissue and smooth muscle. Veins also have **valves** that prevent the **blood** flowing backwards.

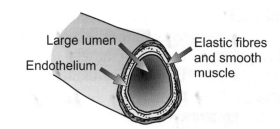

Large lumen

Endothelium

Elastic fibres and smooth muscle

2) When the **leg muscles** contract they bulge and press on the walls of the veins, pushing the blood up the veins. When the muscles relax, the valves close. This action helps the blood **return** to the heart.

HMS Vein — a superior vessel...

1) What is the role of arteries in the circulatory system?

2) Explain the importance of the elastic tissue in the walls of arteries.

3) Describe how arterioles can control the amount of blood flowing to an organ.

4) Capillaries have very thin walls, which sometimes have gaps in them. Explain how these characteristics make capillaries suited to their job.

5) What structure do veins contain, that other blood vessels don't have?

6) Explain how leg muscles help return blood to the heart.

Blood

Haemoglobin has **Special Properties**

1) The blood's main function is to **transport** materials to and from cells.

2) So the blood can do this, red blood cells are packed with **haemoglobin**, a protein that contains iron and can **carry oxygen**.

3) When oxygen combines with haemoglobin it forms **oxyhaemoglobin**.

4) When there's a lot of oxygen present, one molecule of haemoglobin can combine with **four** molecules of oxygen — the haemoglobin is **100% saturated**.

5) When less oxygen is present, fewer molecules of oxygen combine and the haemoglobin is **less than** 100% saturated.

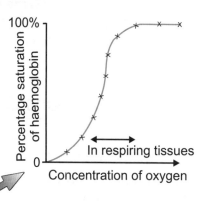

> It would be reasonable to expect that a graph of '% saturation of haemoglobin' against 'concentration of oxygen' would be a straight line (i.e. that the two would be proportional). However, when experiments are carried out and the results plotted, the line of best fit is **S-shaped**.

6) Haemoglobin has special properties that allow it to become fully saturated with oxygen in the capillaries around the **alveoli** of the lungs, where there's a **high concentration** of **oxygen**.

7) Then when it reaches respiring tissue, where there's **less oxygen**, it can give up almost all of its oxygen immediately — so the rate of respiration in the tissues isn't slowed down because of an oxygen shortage.

Carbon Dioxide Changes the **Properties** *of* **Haemoglobin**

1) Respiring tissues produce **carbon dioxide**.

2) If there's not a lot of carbon dioxide present, the haemoglobin is **less efficient** at **taking up** oxygen (i.e. it needs to be exposed to more oxygen before it becomes fully saturated).

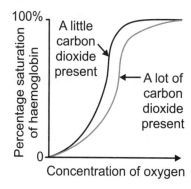

3) But, when there's a lot of carbon dioxide present, the haemoglobin becomes **more efficient** at **releasing** oxygen (i.e. it can release more oxygen molecules in areas of fairly high oxygen demand).

4) This is good because it means that **rapidly respiring tissues**, e.g. contracting leg muscles and brain cells, get **more oxygen**.

5) This effect of carbon dioxide concentration on the oxygen-binding properties of haemoglobin is known as the **Bohr effect**.

<u>The Bore effect — caused by reading this page...</u>

1) Name the substance picked up by the blood in the lungs.

2) How many molecules of oxygen are bound to a haemoglobin molecule when it's fully saturated?

3) Which gas affects the oxygen-binding properties of haemoglobin?

4) Under what circumstances does a tissue require the most oxygen?

Variation and Evolution

We all Vary

1) All organisms are **different** from each other, e.g. giraffes are loads different from zebras, which are different from lions and tigers and bears...

2) Organisms of the **same species** also show **some variation**, e.g. humans show variation in height, weight, favourite colour of shoe polish...

3) Organisms of the **same species** are similar because they all have the **same genes** but they vary because they have **different versions** of those genes (called **alleles**). E.g. humans all have a gene for blood type, but they can have A, B or O alleles.

Variation Means Some Organisms are Better Adapted

1) An adaptation is a **characteristic** that helps an organism to **survive** and **have children**, e.g. polar bears have **thick, white fur** to stay warm and camouflaged in the snow.

2) **Characteristics vary** in a population so some organisms are **better adapted** for certain conditions than others, e.g. polar bears with thicker fur are better adapted to survive in a cold environment than polar bears with thinner fur. The slightly different adaptations you get **within species** (e.g. slightly thicker fur on one polar bear compared to another) are coded for by **different alleles**.

Evolution

1) Evolution is the **gradual change** in the **characteristics** of a population from one generation to the next. The theory of evolution is that all organisms evolved from a **common ancestor** (organism) over **millions of years**.

2) There's **more than one** mechanism by which evolution occurs — one is **natural selection**.

Natural Selection

1) Organisms from the **same population** all **vary** (e.g. different length of fur).

2) Organisms **compete** with each other for food, shelter, water, etc.

3) Those with **better adaptations** (caused by different **alleles**) are more likely to find food, shelter, water, etc., **survive** and have little **kiddies**. So they **pass on** the alleles for their better adaptations. E.g. bears with longer fur will stay warmer and be more likely to survive, and so have kids with longer fur.

4) Over time, the **number** of organisms with the better adaptations (alleles) **increases**.

5) The **whole population** of organisms **evolves** to have the better adaptations (alleles).

Bah, evolution takes ages. I want wings NOW...

1) What is an allele?

2) What is an adaptation?

3) Briefly describe natural selection.

Classification

Classification Systems

1) Classification is just **sorting** organisms into different **groups** and **naming them**.

2) It makes it **easier** for scientists to **study** organisms without getting **confused**, because every type of organism has a different name, e.g. *Homo sapien* (humans) or *Ursus maritimus* (polar bears).

3) Organisms are arranged into different groups depending on their **similarities** and **differences**, e.g. all animals are grouped together, and all plants are grouped together in a separate group because they're different to animals.

4) Organisms are placed in groups in **classification hierarchies** (pronounced: hire-arc-ees) — the biggest groups (e.g. animals, plants) are **split** into **smaller groups** (e.g. animals with a backbone in one group and animals without a backbone in another). These groups are **split again** into more smaller groups, and so on.

As you move down the hierarchy you get <u>more groups</u> at each level but <u>fewer organisms</u> in each group.

First group with the largest number of organisms in them, e.g. the animal kingdom has all animals in it.

The last group is called <u>species</u>. There's only one type of organism in each one and members can have offspring (children) with each other, e.g. humans.

A B   C

Organisms in group A are more similar to organisms in group B than they are to organisms in group C.

A **species** is a group of organisms that **look similar** and can reproduce to give **fertile offspring** (their children can also reproduce).

Classification Systems are Based on Lots of Things

1) **Older** classification systems grouped organisms based only on how they look, e.g. four limbs, six eyes, bum chin...

2) **Newer systems** use looks and lots of other things:
   - **DNA** — how similar and different the base sequence is (e.g. ATTTAC vs. ATTTAT).
   - **Other molecules** — e.g. proteins and enzymes.
   - **Early development** — how they grow from an embryo to a baby.

My poor brother — at least he's not classified with the apes any more...

1) What does classification involve?
2) What is a species?
3) List four things newer classification systems use to group organisms.

Xylem and Phloem

Xylem Tissue Transports Water and Minerals from Roots

Water from the soil **enters** the roots by **osmosis**. Then it travels through the root to the **xylem** — this is the tissue that **transports water** through the plant and up to the leaves.
Water can travel through the roots in **two** ways:

The **symplast system**:
- Some water moves through the root via the **cytoplasm** of the root cells. The water has to cross the **cell membrane**, which regulates the passage of the water and dissolved minerals.

The **apoplast system**:
- The water moves through the **cell walls** and the **spaces between the cells**.
- There are **no membranes** to regulate the passage.

Water Travels Up the Plant Through the Xylem Tissue

The cells that make up the tubes (vessels) of **xylem tissue** are dead, waterproof and hollow. This means water can **move** through them easily. Water is **pulled up** through the xylem tissue by a combination of factors: **cohesion**, **tension** and **adhesion**:

1) Water **evaporates** from inside the leaf leaving a higher concentration of solutes.

2) Water from the nearest xylem vessel enters by **osmosis**.

3) Water molecules stick together because of weak hydrogen bonds between them — this is called **cohesion**.

4) As water molecules leave the xylem vessel they **pull up** further molecules, so the whole column of water is pulled up.

5) Evaporation pulls the water column upwards and gravity pulls it down, so the water column is under **tension**.

6) The **adhesion** of water molecules to the sides of the xylem vessels stops the column breaking.

Phloem Transports Organic Compounds

Sugars and other organic compounds are **transported** through plants in **phloem tissue**. Phloem tissue is also arranged in **tubes** so the solutions of sugar, etc. can **move** through them easily.

1) The movement of carbohydrates and other organic compounds in plants is known as **translocation**.

2) It occurs in the **sieve tubes** of the **phloem tissue**.

3) **Companion cells** next to the sieve tubes are believed to **actively transport** sugar into the sieve tubes, and then water follows by **osmosis**.

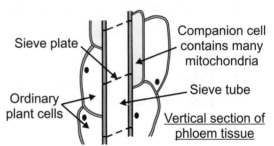

Sieve plate / Companion cell contains many mitochondria / Ordinary plant cells / Sieve tube / Vertical section of phloem tissue

Relax, sit back and just go with the phloem...

1) In the symplast system, which part of the cell does water move through?

2) Why is the column of water in the xylem under tension?

3) What substances are transported in the phloem tissue?

Planning an Experiment

A *Good Experiment* Gives *Precise* and *Valid Results*

1) **Precise** results are **repeatable** (if the same person repeats the experiment using the same methods and equipment, they will get the same results) and **reproducible** (if someone different does the experiment, or a slightly different method or piece of equipment is used, the results will still be the same).

2) **Valid** results are **precise** and **answer the original question**. To get valid results you need to **control all the variables** to make sure you're only testing the thing you want to.

To Get *Good Results* You Need to *Design* Your *Experiment Well*

Here are some of the things you need to consider when thinking about **experimental design**:

1) **Only one variable should be changed** — Variables are **quantities** that have the **potential to change**, e.g. pH. In an experiment you usually **change one variable** and **measure its effect** on another variable.
   - The variable that you **change** is called the **independent variable**.
   - The variable that you **measure** is called the **dependent variable**.

2) **All the other variables should be controlled** — When you're investigating a variable you need to keep everything else that could affect it **constant**. This means you can be sure that **only** your **independent** variable is **affecting** the thing you're measuring (the dependent variable).

3) **Negative controls should be used** — Negative controls are used to **check** that only the independent variable is affecting the dependent variable. Negative controls **aren't expected** to have **any effect** on the experiment.

4) **Repeat the experiment at least three times** — Doing **repeats** and getting **similar results** each time shows that your data is **repeatable**. This makes it more likely that the same results could be **reproduced** by another scientist in an independent experiment. This makes your data **more precise**. Doing repeats also makes it easier to spot any **anomalous results** — unexpected results that don't fit in with the rest.

EXAMPLE: Investigating the effect of **temperature** on **enzyme activity**.

1) Temperature is the **independent** variable.
2) Enzyme activity is the **dependent** variable.
3) pH, volume, substrate concentration and enzyme concentration should all stay the **same**.
4) The experiment should be **repeated** at least three times at each temperature used.
5) A **negative control**, containing everything used except the enzyme, should be measured at each temperature. No enzyme activity should be seen with these controls.

Graphs

You Can Use *Scatter Graphs* to *Present* Your *Data*

1) When you want to show how **two variables** are **related** (or **correlated**, see next page) you can use a **scatter graph**.

2) Make sure that:

- The **dependent variable** goes on the **y-axis** (the vertical axis) and the **independent** on the **x-axis** (the horizontal axis).

- You always **label** the **axes**, include the quantity and **units**, and choose a **sensible scale**.

3) When you draw a **line** (or curve) **of best fit** on a **scatter graph**, draw the line through or as near to as many points as possible, **ignoring** any **anomalous** results.

Scatter graph

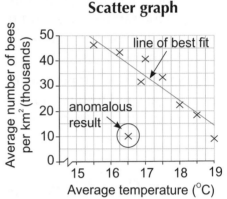

Find the *Rate* By Finding the *Gradient*

Rate is a measure of how much something is **changing over time**. Calculating a rate can be useful when **analysing** your data, e.g. you might want to the find the **rate of a reaction**. Rates are easy to work out from a **graph**.

For a **linear** graph you can calculate the **rate** by finding the **gradient of the line**:

EXAMPLE:

$cm^3 s^{-1}$ means the same as cm^3/s (centimetres per second)

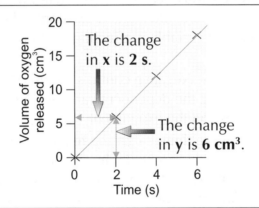

The change in **x** is **2 s**.

The change in **y** is **6 cm³**.

$$gradient = \frac{change\ in\ Y}{change\ in\ X}$$

So in this **example**:

$$rate = \frac{6\ cm^3}{2\ s} = 3\ cm^3\ s^{-1}$$

For a **curved** (non-linear) graph you can find the **rate** by drawing a **tangent**:

EXAMPLE:

1) Position a ruler on the graph at the **point** where you want to know the **rate**.

2) **Angle** the **ruler** so there is **equal space** between the **ruler** and the **curve** on **either** side of the point.

3) **Draw** a **line** along the ruler to make the tangent.

4) **Calculate** the **gradient** of the **tangent** to find the **rate**.

gradient = 55 m² ÷ 4.4 years = **12.5 m² year⁻¹**

Extend the line right across the graph — it'll help to make your **gradient calculation easier** as you'll have **more points** to choose from.

The change in y is **55 m²**.

The change in x is **4.4 years**.

Correlation and Cause

Lines of Best Fit Are Used to Show Trends

The line of best fit on this graph shows that as one variable **increases**, the other variable **also increases**. This is called a **positive correlation**. The data points are all quite close to the line of best fit, so you can say the correlation is **strong**. If they were more spread out, the correlation would be **weak**.

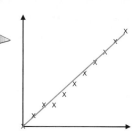

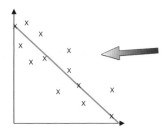

Variables can also be **negatively correlated** — this means one variable **increases** as the other one **decreases**. Look at the way the line of best fit **slopes** to work out what sort of correlation your graph shows.

Sometimes the graph won't show any clear trend and you won't be able to draw a line of best fit. In this case, you say there's **no correlation** between the variables.

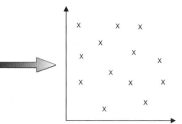

Correlation Doesn't Always Mean Cause

1) Be careful what you **conclude** from an experiment — just because two variables are correlated, it doesn't necessarily mean that one **causes** the other.

2) In lab-based experiments, you can say that the independent variable causes the dependent variable to change — the increase in temperature **causes** an increase in the rate of the reaction. You can say this because everything else has **stayed the same** — nothing else could be causing the change.

3) Outside a lab, it can be much harder:

> **EXAMPLE:**
>
> Kate measured the level of air pollution and the incidence of TB, to see whether the two are related. Her results show a positive correlation between the variables — where the level of pollution is highest, the incidence of TB is also highest.

From Kate's results, you can't say that air pollution causes TB.

Neither can you say that TB causes air pollution.

It could be either way round... or one change might not cause the other at all — you just can't tell.

Answers

Section 1 — Biological Molecules

Page 1 — Proteins

1   The order in which the amino acids are arranged in a protein chain.
2   Hydrogen bonds (weak forces of chemical attraction).

Page 2 — Carbohydrates

1   E.g. glucose and fructose.
2   Maltose
3   Any two from: e.g. starch, cellulose, amylose and amylopectin.

Page 3 — Lipids

1   Carbon, hydrogen, and oxygen.
2   Every carbon atom in saturated fatty acid chains is joined by a single bond. Unsaturated fatty acids have one or more double bonds in their carbon chains.
3   Triglycerides have three fatty acid chains attached to a glycerol molecule. Phospholipids have two fatty acid chains and a phosphate group attached to a glycerol molecule.

Page 4 — Enzymes

1   Increase the rate of/speed up biochemical reactions. / Act as biological catalysts.
2   Activation energy is the energy required to get a reaction started.
3   Break down food into smaller molecules.

Page 5 — Enzymes

1   The substrate molecule must be the correct shape to fit into the active site.
2   The shape of the active site has changed and the substrate will no longer fit.
3   Acids and alkalis can denature enzymes by disrupting the weak hydrogen bonds in the enzymes' tertiary structure and changing the shape of their active sites.

Section 2 — Cell Structure

Page 6 — Eukaryotic and Prokaryotic Cells

1   E.g. a bacterium.
2   E.g. nucleus, cytoplasm, cell-surface membrane and mitochondria.
3   They provide a source of energy for the cell.

Page 7 — Microscopes

1   Any 3 from: nucleus, cell membrane, cytoplasm, mitochondria.
2   An electron microscope.
3   An electron micrograph.

Page 8 — Functions of the Nucleus, Mitochondria and Cell Wall

1   The nucleus.
2   The mitochondria.
3   There is a smooth outer membrane and a folded inner membrane.
4   glucose + oxygen → carbon dioxide + water + (energy)

5   Adenosine triphosphate/ATP.
6   Cellulose

Page 10 — Cell Membranes

1   Phospholipids and protein molecules.
2   Diffusion, osmosis, facilitated diffusion and active transport.
3   Osmosis
4   Water potential
5   Carrier and channel proteins.
6   To provide the energy needed for the process.

Section 3 — Genetics and Cell Division

Page 11 — DNA and Protein Synthesis

1   Double helix
2   Four
3   Adenine, cytosine, guanine and thymine.
4   Each base forms hydrogen bonds to a base on the other strand.
5   This is when pairs of complementary bases join together. A pairs with T and C pairs with G.
6   A section of DNA that codes for a particular protein.
7   The order of bases in a gene.

Page 12 — RNA and Protein Synthesis

1   Because the DNA molecule containing the gene is in the nucleus and is too big to leave the nucleus. But protein synthesis takes place in the cytoplasm, so a copy of the gene that is smaller and can leave the nucleus needs to be made.
2   Messenger
3   Uracil (U)
4   UAACGCGU

Page 13 — Mutations

1   Three (three bases are called a triplet).
2   Changes to the base sequence of DNA.
3   Increase the frequency of mutations.

Page 14 — Chromosomes

1   In the nucleus.
2   23
3   No. They're the same size and carry the same genes but they usually have different alleles.
4   An identical copy of a chromosome.
5   Centromere

Page 15 — Cell Division — Mitosis

1   Growth, repair, asexual reproduction.
2   So each new cell has a full copy of DNA.
3   No (the chromatids separate).
4   Two

Page 16 — Cell Division — Meiosis

1   Haploid
2   In the testes and the ovaries.
3   Two

Answers

4   Four

Section 4 — Exchange
Page 17 — Size and Surface Area to Volume Ratio
1   A small organism.
2   7.5 : 1 or 7.5 to 1
3   Animal A
4   E.g. oxygen, waste products, nutrients.

Page 18 — Structure of the Thorax
1   To provide a large surface area for the efficient diffusion of carbon dioxide and oxygen.
2   The alveoli.
3   The cells are thin and flattened. The gases can diffuse across the cells quickly because the distance is small.
4   Epithelial cells
5 a) There is a higher concentration of oxygen inside the alveoli than in the blood.
   b) Carbon dioxide

Page 19 — Breathing In and Breathing Out
1   When the volume increases, the pressure decreases (and when the volume decreases, the pressure increases).
2   Decrease
3   The intercostal muscles and the diaphragm muscles.

Section 5 — Disease and Immunity
Page 20 — Disease
1   Organisms that can cause disease.
2   E.g. TB, malaria, HIV.
3   Something that increases the chances of something bad happening.
4   Any two from: mouth, lung and throat cancer, emphysema and other lung diseases, cardiovascular disease.

Page 21 — Immunity
1   Foreign antigens.
2   B-cells (B-lymphocytes).
3   To communicate between phagocytes and B-cells / to activate B-cells.
4   Antigens from a pathogen in a form that can't harm you.

Section 6 — The Circulatory System
Page 22 — The Circulatory System
1   The heart.
2   The right and left atria, the right and left ventricles.
3   Arteries, veins and capillaries.
4   Capillaries

Page 24 — The Heart
1   Lungs
2   To keep the blood flowing one way/prevent blood flowing backwards.

3   No
4   Into the ventricles.
5   Because the ventricles have to pump blood further than the atria.
6   To produce the regular electrical impulses that make the atria contract.
7   So it can get oxygen and glucose for respiration (for energy).
8   The coronary arteries.

Page 26 — Blood Vessels
1   They carry blood away from the heart.
2   The elastic stretching and recoil of the artery walls helps to keep blood pressure up and allow important organs to function.
3   When the circular muscle fibres in the walls of arterioles contract, they reduce the size of their lumen. This reduces the amount of blood flowing to the organs that they supply.
4   They allow water and other substances (like glucose and oxygen) to diffuse quickly between the blood in the capillaries and the cells.
5   Valves
6   When leg muscles contract, they push blood up the veins, and when they relax, the valves in the veins close. These actions help push blood back to the heart.

Page 27 — Blood
1   Oxygen
2   Four
3   Carbon dioxide
4   When it's rapidly respiring.

Section 7 — Variation, Evolution and Classification
Page 28 — Variation and Evolution
1   An alternative version of a gene.
2   Any characteristic of an organism that increases its chance of survival.
3   Organisms with good adaptations survive, have children and so pass on the alleles for those good adaptations.

Page 29 — Classification
1   Sorting organisms into groups (based on their similarities and differences) and naming them.
2   A group of organisms that look similar and can reproduce to give fertile offspring.
3   How they look, DNA, other molecules (e.g. proteins and enzymes) and early development.

Section 8 — Plants
Page 30 — Xylem and Phloem
1   Cytoplasm
2   Evaporation is pulling the water column up and gravity is pulling it down.
3   Sugars and other organic compounds.

Index